CLASS 158 AND 159 DMUs

ANDREW COLE

AMBERLEY

First published 2021

Amberley Publishing
The Hill, Stroud
Gloucestershire, GL5 4EP

www.amberley-books.com

ISBN 978 1 4456 8213 6 (print)
ISBN 978 1 4456 8214 3 (ebook)

British Library Cataloguing in Publication Data.
A catalogue record for this book is available from the British Library.

Typesetting by SJmagic DESIGN SERVICES, India.
Printed in Great Britain.

Introduction

The Class 158 units were designed to replace loco-hauled workings, but also to allow a cascade of sprinter units to help eliminate the older, blue square first generation DMUs. They were ordered for the Regional Railways services, and many were delivered with 'Express' branding applied.

A total of 172 units were ordered initially, with BREL Derby Litchurch Lane being responsible for their construction. The first unit emerged in 1989, and was immediately allocated to Haymarket depot following a period of testing.

All of the units were delivered as two-car sets, with the final Regional Railways set delivered in March 1992. A further ten sets were ordered by West Yorkshire PTE for services in and around Leeds. These were delivered concurrently with the initial Regional Railways sets and were eventually absorbed into the Northern fleet.

Seventeen centre cars were also ordered for sets that operated over Trans Pennine routes, with these being made up to three-car sets in late 1991. There is currently only one originally formed three-car Class 158 set in use – No. 158798, allocated to Great Western Railway – with the others having been reformed or renumbered as Class 159 units. The twenty-two three-car Class 159/0 units followed on from the main batch of Class 158 units. They were built as Class 158 units but they would visit Rosyth Dockyard for internal modifications, before being delivered to Network South East for services out of London Waterloo.

The Class 158 unit was a step up in quality when compared to the sprinter units that had preceded them: they had air conditioning fitted from construction and also had a higher maximum speed, being capable of running at 90 mph compared to 75 mph.

The class could be found working the length and breadth of the country, with sets allocated to Inverness in Scotland, through to Cardiff Canton in Wales. A large batch was also allocated to Norwich Crown Point for services through to Birmingham and Liverpool. A small batch of five units was used by the InterCity sector for use on some Cross Country services.

The class has carried many different liveries since privatisation and has been used by many different companies, and they can still be found working all over the country. In 2006, a batch of eight three-car sets were refurbished for South West Trains. They were renumbered 159101–159108 and retained their original formations.

Also, in 2007, a batch of eleven three-car units were retired from Trans Pennine service and were also taken over by South West Trains. They were refurbished by Wabtec at Doncaster and they were reduced to two-car formations, being numbered 158880–158890, with the former centre cars being transferred to Northern Rail and formed into three-car sets at Neville Hill.

All of the sets have received PRM (people with reduced mobility) modifications, which will see them in service for many years to come. The Class 158s are now concentrated on six different train-operating companies, with Northern being the largest operator, followed by Scotrail, East Midlands Railway, Transport for Wales, Great Western Railway and South West Trains. The Class 159 units have only ever been operated by South West Trains. There are currently plans to replace the class in the Transport for Wales and East Midlands Railways fleet, although they will be transferred for further use. All of the sets are still in service, with just one car from No. 158861 scrapped following collision damage, but a replacement was provided using a spare bodyshell.

BREL received an order from the State Railway of Thailand for twenty Class 158 vehicles. They were delivered in 1990/91 and they were painted in a version of Regional Railways livery. There are twelve driving cars and eight intermediate cars and all are still in service today.

The class has proved to be a valuable asset from their Regional Railways days through to their current workings. They will be in use for many years to come, having been fully refurbished throughout their careers. I hope you enjoy looking through my collection of photos, which show the units in their everyday use and in a variety of different liveries that the class has carried.

No. 158702, 21 September 2006
No. 158702 departs from Edinburgh Waverley carrying First Scotrail livery. The class has been associated with services out of Edinburgh since their introduction into service in early 1990. No. 158702 can be seen carrying the name *BBC Scotland 75 Years* underneath the light cluster, but it has since been removed.

No. 158702, 4 September 2017
No. 158702 stands at Edinburgh Waverley waiting to depart for Tweedbank. The unit now carries Scotrail Saltire livery, as does the rest of the Scottish based fleet. As can be seen, the name has now been removed from the cab front – all the names carried by the Scottish-based Class 158s have been removed. The Tweedbank services are operated under the Borders Rail banner.

No. 158703, 6 July 2019

No. 158703 is seen on the buffer stops at Inverness. The class have been allocated to Inverness since their introduction in 1990 for use on the far north lines to Wick and Thurso, and across to Kyle of Lochalsh.

No. 158706, 31 August 1994

No. 158706 stands at Glasgow Queen Street carrying Regional Railways Express livery. It has already received a different style of front skirt, which was fitted to the Scottish fleet to help with clearing the snow. The class were introduced onto the Edinburgh Waverley to Glasgow Queen Street route to replace the Class 47-hauled push-pull sets.

No. 158707, 18 September 2015

No. 158707 arrives at Glasgow Queen Street carrying former First Scotrail livery but with Abellio Scotrail branding. The change of franchise had occurred earlier in the year, with just a change of logo to start with. No. 158707 carries the name *Far North Line 125th Anniversary*, but it has since been removed.

No. 158708, 29 August 1996

No. 158708 is seen stabled at Perth. The unit carries as-delivered Regional Railways Express livery. This view also shows the first-class end of the unit, as indicated by the yellow stripe along the top of the saloon windows.

No. 158709, 16 October 2019
No. 158709 is seen stabled at Inverness waiting its next turn of duty along with classmate No. 158705. The Scottish-based units have also received different headlight clusters when compared to the rest of the fleet, as well as the different style of front skirt.

No. 158711, 15 August 2018
No. 158711 is seen at Glasgow Queen Street, having arrived with a terminating Scotrail service. As can be seen, this was taken during the time when Queen Street was being rebuilt and a number of platforms were extended to accommodate eight-car Class 385 electric units on the Glasgow Queen Street to Edinburgh Waverley express services.

No. 158712, 18 September 2015
No. 158712 is seen arriving at Haymarket, just outside Edinburgh Waverley, with an Abellio Scotrail working. It is seen still carrying former First Scotrail livery, but all the Scottish fleet of forty units have been fully refurbished and repainted into Scotrail Saltire livery since this image was taken.

No. 158713, 21 September 2006
No. 158713 arrives at Edinburgh Waverley with a terminating First Scotrail service. This unit is seen carrying former Scotrail livery, which was applied when National Express operated the franchise, but First Group took over running it in 2004 and were slow to repaint the fleet.

No. 158713, 17 September 2015
No. 158713 stands at Inverness with a Scotrail working to the far north. The Class 158 units took these services over from Metro-Cammell-built Class 156 units and also from Class 37 diesel-hauled workings.

No. 158714, 24 July 1990
No. 158714 is seen arriving at Leicester station before entering service. It is seen on gauging tests, as evidenced by the video cameras fitted to the front. Also of note is the standard front skirt fitted when built, but these were soon replaced with a more angular design to help clear snow.

No. 158715, 19 September 2015
No. 158715 is seen at Perth with a service to Edinburgh Waverley. The name *Haymarket* can be seen on the cab front; it is an adhesive type rather than a cast plate and it has since been removed.

No. 158716, 18 August 2018
No. 158716 arrives at Dundee with a terminating service from Edinburgh Waverley. It would have just traversed the Tay Bridge, and it is seen coupled to a Class 170 unit. Also note the former Great Western Railway HST set stabled up on the carriage sidings, which was present for staff training before the introduction of the trains in Scotland.

No. 158718, 1 September 1995

No. 158718 is seen making a station call at Perth carrying Regional Railway Express livery. Perth is an important junction station in Scotland, where the lines to Aviemore and Inverness diverge from the lines to Dundee and Aberdeen. There are also lines to Edinburgh and Glasgow to the south.

No. 158719, 20 August 2018

No. 158719 arrives at Haymarket with an Abellio Scotrail working to Edinburgh Waverley. At present the first twenty-five members of the class are allocated to Inverness, with the rest of the Scottish fleet – fifteen units – recently transferring from Haymarket to Corkerhill in Glasgow.

No. 158721, 16 October 2019
No. 158721 is seen at Inverness with a service to Kyle of Lochalsh. The Inverness-based fleet has been fully refurbished internally, with different seat alignments making them better for the scenic lines of the far north. Abellio currently have forty of these units allocated.

No. 158722, 4 September 1990
No. 158722 is seen at Derby having just emerged from the nearby Litchurch Lane works where it was built. It is about to be delivered to Haymarket and, as can be seen, carries the original front skirt.

No. 158722, 19 September 2015
No. 158722 is seen after arriving at Perth with a terminating service. The unit is seen carrying unbranded First Scotrail livery, with the new Abellio logo yet to be applied. As seen here, during this time the first-class section was still retained at this end of the unit.

No. 158727, 8 July 2000
No. 158727 arrives at Edinburgh Waverley with a terminating service. It is seen carrying former National Express Scotrail livery. At the time the class could be seen working Fife Circle services as well as services through to Dunblane, which have now been taken over by Hitachi-built Class 385 electric units.

No. 158727, 16 August 2018

No. 158727 is seen at Edinburgh Waverley with a Borders Rail working through to Tweedbank. This line only reopened in September 2015, with the first service entrusted to Class 158 No. 158701. The line is based on the former Waverley Route through Newcraighall and Gorebridge.

No. 158728, 1 September 1994

No. 158728 is seen stabled at Aberdeen and carrying the Regional Railways Express livery. By this time, the Regional Railways branding had been replaced with Scotrail branding. On this day there were no trains running due to an industrial dispute involving signaling staff.

No. 158728, 21 September 2006
No. 158728 departs from Edinburgh Waverley carrying full First Scotrail livery, including a First logo underneath the secondman's windscreen. Of note is the impressive Balmoral Hotel in the background and also the large number of open-top buses on sightseeing duties.

No. 158729, 4 September 2017
No. 158729 is seen at Glasgow Queen Street. It is standing on one of the platforms that have been extended to accommodate eight-car Class 385 units. This is one of the later units that were, until recently, allocated to Haymarket, but it has since transferred to Corkerhill depot in Glasgow.

No. 158730, 16 August 2018

No. 158730 arrives at Haymarket with a service for Edinburgh Waverley. The Haymarket Class 158 units were refurbished following the Inverness-based units, as such they could be seen carrying former First Scotrail livery for longer.

Nos 158734 and 158727, 4 September 2017

Nos 158734 and 158727 are seen side by side at Haymarket. This station is around 5 minutes outside of Edinburgh Waverley, and is a very busy station, with the depot located just to the west of it. It is also an important interchange point for the Edinburgh tram.

No. 158735, 5 September 2017

No. 158735 is seen passing through Partick in the suburbs of Glasgow with an empty coaching stock move from Corkerhill depot to Glasgow Queen Street station. Diesel units are very rare at Partick, except for ECS moves.

No. 158739, 29 August 1996

No. 158739 is seen making a station call at Perth, still carrying Regional Railways Express livery. This has just come from the Dundee direction and will soon head off towards either Edinburgh Waverley or Glasgow Queen Street.

No. 158743, 28 August 1996

No. 158743 is seen arriving at Inverness with a terminating service. This unit was transferred away to Heaton, Newcastle, in 2000, and would eventually be refurbished for use with South West Trains, gaining the number '158882' in the process.

No. 158745, 10 January 2001

No. 158745 is seen departing from Newport with an Alphaline Wales and Borders working, heading towards Cardiff Central. This unit carries Wales and Borders livery, as indicated by the blue saloon doors. These services were named Alphaline under Regional Railways, and it was a branding used to indicate services operated by higher-quality Class 158 units.

No. 158746, 10 September 2002

No. 158746 is seen at Cardiff Central carrying Alphaline Wessex Trains livery, as indicated by the maroon-coloured saloon doors. This unit was later absorbed into the First Great Western fleet and has since been disbanded, with car No. 52746 now formed in unit No. 158959 and car No. 57746 formed in unit No. 158958. At this time the unit carried the name *Spirit of the South West*.

No. 158746, 16 August 2007

No. 158746 arrives at Bristol Temple Meads carrying Alphaline Wessex Trains livery. The Alphaline branding was dropped after the Class 158s were taken into the First Great Western fleet in 2006, but the company was slow to remove the branding. The livery seen has been slightly modified from the previous view, losing the yellow around the cab and also dropping the Wessex Trains logos.

No. 158747, 16 August 2007

No. 158747 departs from Bristol Temple Meads carrying Alphaline Wessex Trains livery. This unit also carried the name *Richard Trevithick* – the plate can be seen above the third saloon window from the rear of the unit. It also carries advertising vinyl for Richard Trevithick's 200th anniversary of railway achievements.

No. 158749, 13 December 2003

No. 158749 arrives at Westbury carrying Alphaline Wessex Trains livery. This was one of a small batch of five Class 158 units that were used by the InterCity sector for use on some Cross Country services before they passed to Wessex Trains.

No. 158749, 19 April 2008

No. 158749 is seen at Weymouth carrying First Great Western local lines livery. This had recently been repainted, losing its former Alphaline Wessex Trains livery, and is seen heading for Castle Cary. Shortly after this view No. 158749 was disbanded and the two cars were used to make three-car Class 158/9 units, but they have since been reformed to form No. 158749 once more.

No. 158750, 16 August 2007

No. 158750 is seen departing from Bristol Temple Meads. This was the last Class 158 unit to carry some form of Regional Railways Express livery, albeit with the addition of a blue stripe with First Trans Pennine logos applied. It had transferred to the South West during the previous year, but retained this livery combination until it received First Great Western local lines livery.

No. 158751, 26 May 2007
No. 158751 is seen at Plymouth carrying plain silver livery and Alphaline logos, but is missing the white logo that was usually applied over the windows. This was the last member of the class to carry the later style front skirt, with the remaining fleet carrying the as-delivered straight skirt. This unit has since been disbanded and car No. 52751 is in set No. 158951 and car No. 57751 is in set No. 158950.

No. 158752, 22 February 2017
No. 158752 stands at Manchester Victoria with a Northern working through to Leeds. This was the first member of the class to receive the Northern white livery following refurbishment. It now has centre car No. 58716 inserted, which came from No. 158799.

No. 158752, 16 June 2017
No. 158752 is seen at Leeds with a Northern working through to Manchester Victoria. The three-car Northern Class 158s are normally used on these workings, although they do pop up on other services from time to time.

No. 158754, 4 June 2013
No. 158754 stands at York having arrived with a terminating Northern Rail working. This franchise was operated for twelve years and was owned by Serco-Abellio, until it was replaced with the Arriva Trains run Northern franchise. No. 158754 is seen with centre car No. 58708, which came from unit No. 158808, which was refurbished, becoming two-car No. 158889 for South West Trains.

No. 158755, 13 April 1991
No. 158755 is seen stabled at Saltley depot when just a few months old. It was highly unusual to get DMUs on Saltley at this time, although a few Class 158 units did appear on occasion. Since this image was taken No. 158755 has had car No. 58702 inserted and is now a three-car unit.

No. 158755, 23 August 2007
No. 158755 departs from Doncaster while heading for Sheffield. The unit is seen carrying former First North Western stars livery. The franchise had been taken over by Northern Rail in 2004, yet the unit still retained the livery, but with Northern logos applied.

No. 158756, 22 August 2016

No. 158756 arrives at Leeds with a Northern working through to York from Manchester Victoria. This view shows the former Northern Rail livery and shows how the centre car just carried plain blue. The centre car, No. 58712, came from unit No. 158812 in 2008.

No. 158756, 16 June 2017

No. 158756 arrives at Manchester Victoria. By this time the unit had received the new Northern white-based livery and it really suits the class well. Manchester Victoria is a very busy station with local services through to Stalybridge, and also through to Blackburn and Wigan; there are also longer distance Trans Pennine Express services that use the station.

No. 158759, 28 May 1998
No. 158759 passes through Stafford carrying North West Trains stars livery. The class was used on services through to London Euston, although they would only last a couple of years. The company was later brought out by First Group and their own livery would later be applied.

No. 158759, 16 July 2013
No. 158759 is seen waiting for departure at Leeds with a Northern Rail working through to York. By this time No. 158759 had centre car No. 58713 inserted from unit No. 158813. Leeds is another very busy station in the north of England and, with Neville Hill depot having a sizeable fleet of Class 158 units, it is a good place to see them working.

No. 158761, 9 May 1991
No. 158761 passes through Washwood Heath on the outskirts of Birmingham with a working through to Norwich. When introduced, the class replaced Class 156 units on this route, bringing with them a higher top speed as well as a more comfortable environment.

No. 158763, 27 March 1991
No. 158763 arrives at Peterborough with a working to Norwich, having been in service for little over a month. Of note is the centre gangway door, which was Perspex and see-through. These were later replaced by solid doors.

No. 158763, 8 August 2008

No. 158763 is seen making a station call at Salisbury with a First Great Western working from Portsmouth Harbour through to Cardiff Central. These services are incredibly busy, and as a result are now operated by three-car sets – either Class 158s or Class 166s.

No. 158766, 19 April 2008

No. 158766 sits at the buffer stops at Weymouth with a First Great Western working through to Castle Cary. The unit is seen carrying a variety of different liveries, the main being former Northern Spirit purple and large 'N' – this has had a blue stripe added to the bottom with First Trans Pennine branding. The unit actually transferred south two years earlier, but First Great Western was very slow to adopt their livery on these units.

No. 158766, 13 January 2016
No. 158766 is seen departing from Bristol Temple Meads. By this time the unit had received First Great Western local lines livery. This took a very novel approach to livery design as all of the white, blue and magenta lines are made up of place names along the company's routes.

No. 158768, 23 October 1998
No. 158768 arrives at Doncaster carrying Trans Pennine Express Northern Spirit livery. This unit was operated by Arriva Trains Northern at the time. They would be replaced by new Class 185 sets built by Siemens, with the majority of the Class 158s transferring to First Great Western.

No. 158770, 27 May 2016

No. 158770 arrives at Leicester with a terminating East Midlands Trains working from Lincoln. EMT have a sizeable fleet of Class 158 units, mainly operating on the Norwich to Liverpool Lime Street corridor, although units did appear on other services from time to time as can be seen.

No. 158773, 5 January 2018

No. 158773 arrives at Norwich with a terminating East Midlands Trains working from Liverpool Lime Street. This unit carries the name *Eastcroft Depot* – it can be seen halfway along the first carriage. Since this image was taken the name has been removed and the unit now carries plain white livery, as the franchise has now passed from Stagecoach to Abellio.

No. 158773, 12 November 2019
No. 158773 is seen at Norwich waiting to depart for Liverpool Lime Street. The East Midlands franchise has recently changed hands to Abellio. No. 158773 is seen carrying a plain white livery with orange doors until a final version of the new livery is applied.

No. 158774, 19 October 2007
No. 158774 departs from Manchester Piccadilly with a Central Trains working from Liverpool Lime Street to Norwich. These services later passed to East Midlands Trains when Central Trains was disbanded. No. 158774 carries a mix of Northern Spirit purple livery but with the addition of Central Trains branding and logos around the cab front.

No. 158775, 10 September 1991

No. 158775 passes through Washwood Heath on its way to Birmingham New Street from Norwich. Today these services are operated by Cross Country Trains using Class 170 units. No. 158775 would later be refurbished for South West Trains and be renumbered '158885'.

No. 158776, 22 May 2008

No. 158776 departs from Bristol Temple Meads while still carrying former Northern Spirit livery, but with a blue First Trans Pennine stripe added. A number of Class 158 units operated around the First Great Western network at this time in this livery until they received local lines livery. No. 158776 has since been renumbered '158958' and has had car No. 57746 added, making it a three-car set.

No. 158777, 18 August 1999

No. 158777 departs from Leeds carrying Trans Pennine Express Northern Spirit livery. This unit unusually carries its running number on the secondman's windscreen rather than on the driver's side. Today No. 158777 can be seen working for East Midlands Railway, based at Derby Etches Park depot.

No. 158782, 19 September 2015

No. 158782 is seen waiting for departure at Edinburgh Waverley with a Scotrail working to Glenrothes. Since this image was taken No. 158782 has been transferred south to Northern, but following refurbishment still retains the distinctive Scotrail-style light clusters.

No. 158783, 30 August 2005

No. 158783 is seen at Plymouth with a Wessex Trains working but carrying Central Trains livery. At the time National Express operated both franchises and No. 158783 was sent south to help out with capacity issues during the summer months.

No. 158784, 15 June 2017

No. 158784 departs from Leeds with a Northern working through to Manchester Victoria. This unit carries the name *Barbara Castle*, with the plate above the first saloon window, and also advertising vinyl for PTEG (Passenger Transport Executive Group).

No. 158785, 24 June 2008

No. 158785 departs from Manchester Piccadilly with an East Midlands Trains working to Norwich. The unit carries former Central Trains livery but with East Midlands Trains logos now applied and, unusually, the full carriage number is also carried on the cab front – most just carry the Nos 52 or 57 to denote which car is leading.

No. 158785, 5 January 2018

No. 158785 is seen waiting to depart from Norwich with an East Midlands Trains working to Liverpool Lime Street. This is one of the longer rail journeys on the network, taking 5 hours and 35 minutes to complete.

No. 158786, 24 July 1996
No. 158786 passes Washwood Heath on its way to Birmingham New Street. Of note on this working is the fact that the hopper windows have been opened in the leading carriage, indicating a failure of the air conditioning – the windows can only be opened by the guard using a carriage key.

No. 158787, 6 August 2009
No. 158787 arrives at Doncaster with a Northern Rail working from Sheffield. The unit carries Northern Rail livery but with the addition of Metro advertising vinyl, which were applied to a large number of the class.

No. 158788, 24 May 1998
No. 158788 is seen having arrived at Chester with a terminating service from Birmingham New Street. The unit still carries Regional Railways Express livery, but 'Express' has been covered with a Central Trains logo and the small Regional Railways branding removed.

No. 158788, 12 June 2008
No. 158788 arrives at Crewe with an East Midlands Trains working from Derby. When the Central Trains franchise was disbanded most of the Class 158 units passed to East Midlands Trains, who applied their logos over the top of the former Central Trains green livery.

No. 158789, 26 July 1996
No. 158789 departs from Tamworth while heading for Birmingham New Street. This unit can be seen, unusually, carrying its running number in black underneath the light cluster, rather than in the normal white on the driver's windscreen.

No. 158789, 20 August 2018
No. 158789 arrives at Haymarket with a Scotrail working to Edinburgh Waverley. This view shows signs of collision damage with part of the front skirt missing. Since this image was taken No. 158789 has transferred south to become part of the Northern fleet.

Nos 158790 and 165013, 30 October 1991
Nos 158790 and 165013 are seen side by side at Derby. No. 165013 was on its delivery run from York works down to Aylesbury to start service on the Chiltern lines out of London Marylebone. Today No. 158790 can be seen working for Northern and is based at Neville Hill, Leeds.

No. 158791, 8 August 2005
No. 158791 departs from Doncaster heading for Sheffield with a Northern Rail working. The unit carries unbranded Central Trains livery, with it having transferred north the previous year.

No. 158791, 22 February 2017

No. 158791 is seen waiting for departure at Leeds with a Northern working through to Nottingham. The unit carries the name *County of Nottinghamshire* and still carries former Northern Rail livery. It will later be refurbished and gain the new Northern white livery.

No. 158792, 16 January 2007

No. 158792 is seen having arrived at York with a Northern Rail working from Selby. The unit still carries full Central Trains livery, including logos, but had been transferred north from Tyseley the previous August.

No. 158793, 4 June 2013

No. 158793 is seen in one of the south-facing bay platforms at York, having arrived with a terminating Northern Rail working. During its time with Central Trains No. 158793 was one of the sets disbanded to make up three-car sets, with car No. 52793 being in set No. 158954 with No. 158786, and car No. 57793 being in set No. 158955 with No. 158794.

No. 158794, 15 June 2017

No. 158794 arrives at Leeds carrying Northern Rail livery with Metro advertising vinyl. It also still retains its Remembrance Day poppy stickers by the front passenger door. This unit has also carried the numbers '158957' and '158955' at different times during its service with Central Trains with the addition of cars Nos 52789 and 57793 respectively.

No. 158798, 13 January 2016

No. 158798 arrives at Bristol Temple Meads with a Great Western Railway working to Cardiff Central. This unit carries overall advertising for the Springboard Opportunity Group and is the only member of the original three-car sets to still retain its original formation as a Class 158, with several others being refurbished as Class 159/1 units.

No. 158798, 26 August 2018

No. 158798 stands at Cardiff Central after having arrived with a terminating Great Western Railway working from Portsmouth Harbour. This set has centre car No. 58715 running in its formation, and has done since it was built.

No. 158799, 13 January 1993

No. 158799 is seen in one of the north-facing bays at Crewe station, formed as a three-car set with car No. 58716 in the centre. Since this view the centre car has transferred to unit No. 158752 and the two remaining cars from No. 158799 are now operated by East Midlands Railway.

No. 158799, 15 April 2016

No. 158799 arrives at Nottingham with an East Midlands Trains working to Leicester. This unit used to run as a three-car set for use by Trans Pennine Express, but the centre car was removed and placed in set No. 158752.

No. 158800, 27 August 1997
No. 158800 is seen departing southwards from Preston. The three-car Class 158 units were originally delivered as two-car sets. No. 158800 would have car No. 58717 added after six months of service, in December 1991. In this view the centre car has been temporarily removed.

No. 158800, 6 April 1998
No. 158800 speeds northwards through Thirsk. As can be seen, the centre car had been reformed into the unit by this time. No. 158800 would later be refurbished for South West Trains, becoming No. 159101, while still retaining its original formation.

No. 158802, 10 August 1999
No. 158802 is also seen at Thirsk, but this time the unit carries Trans Pennine Express Northern Spirit livery. This would be another unit refurbished for South West Trains, but it would become No. 158888 and the centre car (No. 58702) would be moved to unit No. 158755.

No. 158803, 10 May 1992
No. 158803 is seen at Manchester Victoria running as a three-car set. Of note is the fact that some of the three-car sets for use on Trans Pennine workings had the same front skirt fitted as the Scottish sets had, with it being more angular to help with the snowy conditions.

No. 158803, 29 August 2006

No. 158803 is seen in the yard of Wabtec at Doncaster, partway through a very thorough refurbishment for South West Trains. The unit has been totally gutted, as can be seen, and when completed it would emerge carrying the number '159102'.

No. 158805, 23 May 1996

No. 158805 arrives at Liverpool Lime Street with a terminating service from Newcastle Central. As seen here, this unit has lost its running number from the cab front – quite a few of the Trans Pennine units had this problem. No. 158805 was later refurbished for South West Trains, becoming No. 159104 in the process.

No. 158806, 22 October 1996
No. 158806 departs from the magnificent station at Darlington with a southbound Trans Pennine working. The Class 158 units would later be ousted from these services by Class 185 units, with No. 158806 being reduced to a two-car set and passing to Central Trains, while the centre car – No. 58706 – found its way into set No. 158757.

No. 158806, 16 January 2007
No. 158806 is seen departing from York. The unit now carries former Northern Spirit livery but with the blue First Trans Pennine stripe added. Also of note is the original-style front skirt that is fitted – not all of the Trans Pennine sets received the angular skirts.

No. 158808, 7 September 1991

No. 158808 arrives at Leeds with a working through to Newcastle. The unit is running as a two-car set; it only ran like this for the first two months of service before car No. 58708 was added. No. 158808 would later be refurbished, becoming No. 158889 in the process, and the centre car would be reformed into No. 158754.

No. 158810, 19 October 2007

No. 158810 departs from Manchester Piccadilly with a Central Trains working from Liverpool Lime Street through to Norwich. By this time the centre car – No. 58710 – had been reformed into No. 158753 and the unit carries Central Trains logos on top of the former Northern Spirit livery.

No. 158812, 2 August 2007
No. 158812 departs from Peterborough with a Central Trains working through to Norwich. This was another set transferred from Trans Pennine Express in 2007, and it too gained the Central Trains logos on top of the old Northern Spirit livery. The centre car (No. 58712) had been reformed into unit No. 158756 by this time.

No. 158812, 24 June 2008
No. 158812 arrives at Manchester Piccadilly with an East Midlands Trains working to Liverpool Lime Street. By this time the unit is seen carrying liveries and logos of three different operators, with East Midlands Trains branding applied to the former Northern Spirit livery. The former Central Trains logos can also be seen around the front cab.

No. 158813, 16 January 2007

No. 158813 stands at York in the last days of its service with First Trans Pennine; it would pass later in the year to Central Trains. The centre car (No. 58713) would be reformed into set No. 158759 before its transfer.

No. 158815, 13 May 2002

No. 158815 is seen stabled most unusually on Saltley depot, Birmingham. The unit still retains its Regional Railways Express livery but has now received Alphaline Wales and Borders Trains logos.

No. 158816, 1 August 2009

No. 158816 is seen on the buffer stops at Leeds with a Northern Rail working through to Nottingham. These services are due to go over to Class 195 operation, freeing up the Class 158 units to help eliminate the Class 142 Pacer units from Northern.

No. 158817, 30 December 2016

No. 158817 is seen at Manchester Victoria after having arrived with a Northern working from Leeds. This unit was sent new to Cardiff Canton and worked in Wales for the first sixteen years of service before transferring northwards in 2007.

No. 158820, 17 May 1998
No. 158820 arrives at Westbury carrying Regional Railways Express livery, but with just a plain Alphaline logo applied. This unit has spent its whole service life allocated to Welsh depots, firstly Cardiff Canton, then at Machynlleth.

No. 158822, 13 August 2013
No. 158822 is seen arriving at Cardiff Central carrying Arriva Trains Wales livery. The Welsh-based Class 158 units are predominantly used on services through Shrewsbury to Birmingham International, and to Aberystwyth and up the Cambrian Coast to Pwllheli.

No. 158822, 23 May 2019

No. 158822 passes through Marston Green on the outskirts of Birmingham with a Transport for Wales (TfW) service through to Aberystwyth. TfW took over the running of the Welsh franchise in October 2018, introducing a new white and red livery in the process.

No. 158823, 12 June 2008

No. 158823 is seen waiting for departure time at Manchester Piccadilly. It is seen carrying the original Arriva Trains Wales livery of turquoise and white, which would later be replaced with a darker blue version.

No. 158823, 16 July 2013

No. 158823 is seen departing from Crewe with an Arriva Trains Wales working through to Birmingham International. On its return working to Aberystwyth it would collide with a tractor at Buttington, resulting in it being out of traffic for nearly a year receiving repairs.

No. 158825, 30 August 2005

No. 158825 is seen at Plymouth carrying advertising livery for Ginsters. Five Class 158 units received this livery – Nos 158819, 821, 825, 827 and 841 – and they carried it for a number of years.

No. 158827, 17 August 1998
No. 158827 is seen waiting to depart from Paignton. This is the terminus station on the line from Newton Abbott and it's also home to the Dartmouth Steam Railway, whose line is to the right of the main platforms.

No. 158828, 13 August 2013
No. 158828 is seen making a station call at Cardiff Central carrying Arriva Trains Wales livery. There are currently twenty-four Class 158 units in the Welsh fleet, but they are now operated by Transport for Wales, who have set Nos 158818 to 841 available for service.

No. 158831, 1 April 2002

No. 158831 stands at Plymouth carrying Alphaline Wales and Borders livery. It would later work a service down to Penzance, and today these services are run by Great Western Railway using Class 150 units.

No. 158831, 14 December 2016

No. 158831 arrives at Cardiff Central carrying Arriva Trains Wales blue livery, complete with bilingual logos. This livery is fast disappearing in favour of Transport for Wales white and red.

No. 158833, 4 August 2018

No. 158833 arrives at a sunny Cardiff Central carrying Arriva Trains Wales livery. Back in 1994 this unit was involved in a serious collision with Class 43 No. 43071 at Newton Abbott, resulting in the one car being sent to Doncaster for repairs, which took three months to complete.

No. 158836, 8 May 2012

No. 158836 is seen at Birmingham International, waiting to depart with the 12.09 Arriva Trains Wales service to Aberystwyth, coupled with classmate No. 158831. The unit carries former Alphaline Wales and Borders livery but with Arriva logos applied over the plain silver.

No. 158840, 1 April 2008

No. 158840 departs from Manchester Piccadilly carrying the original Arriva Trains Wales livery of turquoise with cream doors while heading for Llandudno.

No. 158841, 21 May 2019

No. 158841 arrives at Manchester Piccadilly carrying Transport for Wales livery. TfW took over the running of the Welsh franchise in 2018 and has already ordered replacement units from Stadler, which will allow their fleet of Class 158 units to be withdrawn from service and used by other train operating companies.

No. 158843, 24 June 2008

No. 158843 is seen at Manchester Victoria with a Northern working through to Leeds. This unit had previously seen use with Alphaline Wales and Borders and still retained their base livery, but with small Northern branding applied.

No. 158844, 22 August 2016

No. 158844 is seen after arriving at Leeds with a terminating Northern working. The unit is seen carrying 'We are Northern' branding, which was applied to quite a few members of the Class 158 fleet at the time.

No. 158845, 22 February 2017
No. 158845 waits for its departure time at Leeds with a Northern working through to York. This is another member of the class with 'We are Northern' vinyl applied. This unit was part of a batch that was allocated new to Norwich for services through to Birmingham and Liverpool.

No. 158846, 2 August 2007
No. 158846 arrives at Peterborough with a Central Trains working to Liverpool Lime Street. When Central Trains was disbanded these services passed to East Midlands Trains, along with the Class 158 units.

No. 158846, 12 June 2008
No. 158846 is seen arriving at Manchester Piccadilly, this time after the end of the Central Trains franchise. It is seen with their base livery, but with East Midlands Trains logos applied. The East Midlands Class 158s are due to be replaced with inherited Class 170 units in the near future at the time of writing.

No. 158847, 24 July 2017
No. 158847 stands under the impressive train shed at Liverpool Lime Street waiting to depart for Norwich with classmate No. 158780. The two sets will be split at Nottingham, with just a single two-car set carrying on to Norwich. No. 158847 can be seen carrying the name *Lincoln Castle Explorer*.

No. 158848, 13 June 1995
No. 158848 approaches Water Orton on the outskirts of Birmingham carrying as-built Regional Railways livery. Of note is the absence of the Express logos that were applied to earlier class members.

No. 158848, 23 August 2007
No. 158848 departs from Doncaster with a Northern working to Sheffield. The unit carries former Central Trains livery but with Northern branding applied. This finished service with Central Trains in December 2006 and was put into storage at Eastleigh, along with quite a few other members of the class, before being reinstated by Northern four months later.

No. 158849, 10 March 2015

No. 158849 departs from Leeds with a Northern working through to Blackpool North. This unit carries a special advertising livery in connection with the 2014 Tour de France cycle ride, which had included stages in Leeds and York.

No. 158849, 21 May 2019

No. 158849 stands at Leeds after having arrived with a terminating service from Manchester Victoria. By this time the unit had been fully refurbished internally and externally, bringing it up to date with today's PRM requirements.

No. 158850, 1 August 2009
No. 158850 is seen stabled at Carlisle station waiting its next turn of service. This was one of eleven members of the class stored at Eastleigh Works at the end of 2006 after they had finished service with Central Trains, before they were all reinstated with Northern.

No. 158854, 1 April 2008
No. 158854 departs from Manchester Piccadilly with an East Midlands Trains working to Norwich. The unit carries former Central Trains livery but with no other branding, despite East Midlands operating the unit for the past six months. This unit has since been named *The Station Volunteer*.

No. 158855, 10 March 1996
No. 158855 is seen stabled at its then home depot of Norwich Crown Point. This would spend six years allocated to NC before it was transferred to Tyseley. This was later transferred to Wessex Trains and received the name *The Exmoor Explorer*, but it later transferred to East Midlands Trains and lost the name in the process.

No. 158859, 5 October 2018
No. 158859 is seen approaching Doncaster from the Sheffield direction with a Northern working to Cleethorpes. Today these services have been taken over by former Scotrail Class 170 units, with the Class 158s transferring up to Heaton to help eliminate the Class 142 Pacer units from the north-east.

No. 158861, 7 February 2009
No. 158861 approaches Carlisle with a terminating Northern Rail working from Newcastle. This unit was involved in a serious collision when just one month old at Stockport in 1992, when it collided with Class 47 No. 47343, resulting in car No. 52861 being written off. A spare bodyshell was used to create a 'new' No. 52861 and the set was out of use for twelve months. This unit has carried two different names during its career, *Spirit of the South West* and *Magna Carta 800 Lincoln 2015*.

No. 158864, 24 June 2008
No. 158864 departs from Manchester Piccadilly carrying former Alphaline Wessex Trains livery but with East Midlands Trains logos applied.

No. 158866, 16 August 2007
No. 158866 is seen stabled among the weeds at Bristol Temple Meads station carrying full Alphaline Wessex Trains livery. There were twelve members of the class in the Wessex Trains fleet at the time. They were eventually split between First Great Western and East Midlands Trains when the Wessex franchise ended.

No. 158867, 31 August 1999
No. 158867 departs from Newport and is heading towards Cardiff Central. This unit is carrying Alphaline Wales and West livery and was the only member of the class to do so. It was later transferred to First Scotrail.

Nos 158867 and 158869, 18 September 2015
Nos 158867 and 158869 are seen side by side at Haymarket, both carrying Scotrail Saltire livery. These two units have led remarkably similar lives: both were delivered new to Cardiff before being transferred to Haymarket in 2007, they have both since moved to Northern, and both have been fully refurbished.

No. 158869, 4 June 1992
No. 158869 is seen passing through Washwood Heath in Birmingham on its delivery run from Litchurch Lane Derby to Cardiff Canton. It is seen with classmates Nos 158870 and 158871, which at the time left just one more unit to be delivered to Regional Railways.

No. 158869, 17 August 2018
No. 158869 is seen twenty-six years after the previous view, and by this time it had been transferred to Scotrail. It is seen at Perth waiting to depart for Edinburgh Waverley. It has since moved to Northern.

No. 158870, 21 May 2019
No. 158870 departs from Manchester Piccadilly carrying unbranded Scotrail Saltire livery. This was one of eight Class 158 units that transferred south from Scotland following the introduction of Class 385 electric units north of the border.

No. 158870, 3 October 2019
No. 158870 is seen at LNWR Crewe having been fully refurbished for Northern. Of note is that the unit retains several former Scottish features, including the angled front skirt and the unusual headlight clusters fitted to the Scottish examples.

No. 158872, 16 August 2007
No. 158872 departs from Bristol Temple Meads carrying Alphaline Wessex Trains livery. This was the last of the Regional Railways Class 158s delivered, with just the Class 159 units to follow.

No. 158880, 8 August 2008
No. 158880 is seen arriving at Eastleigh carrying South West Trains livery. This was the first of eleven Class 158 units that were refurbished and renumbered for SWT use. This unit was originally numbered '158737' and was initially based at Haymarket.

No. 158881, 24 June 2008
No. 158881 arrives at Crewe with an East Midlands Trains working from Derby. This was a South West Trains unit at the time and was on hire to East Midlands Trains. It was originally numbered '158742' and was refurbished in 2007.

No. 158881, 2 June 2018

No. 158881 departs from Eastleigh while heading towards Romsey. This unit, along with the majority of the former South West Trains units, is now operated by South Western Railway and in this view the unit carries their logos on top of former South West Trains livery.

No. 158882, 24 June 2008

No. 158882 is seen departing from Manchester Piccadilly while on hire from South West Trains to East Midlands Trains. Both companies were operated by the Stagecoach Group at the time, therefore it was easy to hire units from one company to another. No. 158882 was renumbered from Class 158 No. 158743 when it was refurbished for South West Trains.

No. 158882, 26 April 2017
No. 158882 is seen at Longbridge on the outskirts of Birmingham while on its way back to Salisbury following refurbishment at Brush, Loughborough. The unit carries full South West Trains livery, despite the franchise only having another four months to run.

No. 158883, 23 May 2009
No. 158883 departs from Romsey carrying South West Trains livery. This was renumbered from Class 158 No. 158744 and was originally based at Haymarket for the first ten years of its working life.

No. 158884, 3 November 2017

No. 158884 is seen waiting to depart from Bristol Temple Meads. It is seen, most unusually, coupled with Great Western Railway Class 153 No. 153305. No. 158884 was refurbished and renumbered from Class 158 No. 158772.

No. 158885, 23 May 2009

No. 158885 is seen at Romsey with a South West Trains working through to Salisbury. This was refurbished from Class 158 No. 158775 and spent time allocated to Norwich Crown Point and Heaton in its early career.

No. 158886, 4 October 2019
No. 158886 is seen approaching Eastleigh from the Romsey direction. This unit carries South Western Railway logos on top of South West Trains livery. No. 158886 was refurbished and renumbered from Class 158 No. 158779.

No. 158887, 1 April 2008
No. 158887 is seen at Nottingham while on hire from South West Trains to East Midlands Trains. The unit was on hire for around three months before returning south. It was refurbished from Class 158 No. 158781.

No. 158888, 22 May 2008
No. 158888 is seen arriving at Salisbury carrying South West Trains livery. This was renumbered from Class 158 No. 158802 and it lost its former centre car in the process, with Class 158 No. 158755 gaining car No. 58702.

No. 158888, 5 October 2019
No. 158888 makes a station call at Eastleigh carrying South Western Railway livery. Most of the SWR Class 158/8 units now carry the new livery.

No. 158889, 22 May 2008

No. 158889 is seen stabled at Salisbury station. This was renumbered from Class 158 No. 158808, with its centre car finding its way into Class 158 No. 158754. This would later move permanently to East Midlands Trains.

No. 158889, 6 September 2019

No. 158889 is seen at Sheffield with an East Midlands Railway working to Liverpool Lime Street. The unit is seen carrying EMR Regional logos applied over the top of its former East Midlands Trains livery.

No. 158901, 20 June 2000
No. 158901 passes through Thirsk while heading northwards. The Class 158/9 units were ordered by West Yorkshire PTE and carried West Yorkshire red and cream livery with MetroTrain logos, in which No. 158901 is seen.

No. 158901, 15 April 2016
No. 158901 is seen at Nottingham waiting to depart with a Northern Rail working to Leeds. By now the unit carries Northern Rail livery but also has advertising vinyl for Calder Valley.

No. 158902, 24 October 1996
No. 158902 arrives at Leeds while carrying West Yorkshire MetroTrain livery. The ten Class 158/9 units have always been allocated to Neville Hill depot for services in and around Leeds. Note the similar-liveried Class 308 electric unit alongside.

No. 158902, 9 May 2006
No. 158902 is seen departing from Doncaster while heading for Sheffield. By this time the unit had received the Northern version of Metro livery, which was a bold change from the original West Yorkshire livery.

No. 158904, 18 October 2019

No. 158904 is seen arriving at Carlisle with a Northern working from Leeds. This unit is seen carrying Northern white livery as the units are now used in a common pool. The white livery indicates that the unit has been fully refurbished – both internally and externally.

No. 158905, 20 March 2003

No. 158905 is seen departing from Doncaster heading for Sheffield. It is seen carrying the original West Yorkshire Metro livery, but the British Rail double-arrow symbol underneath the first saloon window has been replaced with an Arriva logo.

No. 158906, 22 February 2017
No. 158906 is seen at Leeds carrying Northern Rail livery but with the addition of Calder Valley adverts. It also has stickers for the free Wi-Fi that had been installed and also USB ports. Also note the Remembrance poppy stickers underneath the first saloon window.

No. 158907, 21 May 2019
No. 158907 arrives at Manchester Piccadilly with a Northern working through to Manchester Airport. All of the Northern Class 158/9 units carried Calder Valley adverts on their Northern Rail liveries.

No. 158908, 3 July 2019

No. 158908 is seen poking out of the LNWR depot at Crewe. This facility was responsible for some of the refurbishments on the Northern Class 158 fleet, with the work taking a couple of weeks. No. 158908 would be released back to traffic carrying Northern white livery.

No. 158909, 22 April 1996

No. 158909 is seen arriving at Liverpool Lime Street, having come across the Pennines from Leeds. It carries the original West Yorkshire Metro livery.

No. 158909, 23 October 1996
No. 158909 departs from York carrying original West Yorkshire Metro livery. The final ten members of the class were given their own number series as they were ordered by West Yorkshire PTE.

Nos 158950 and 175103, 13 August 2013
Nos 158950 and 175103 are seen waiting to depart from Cardiff Central with a First Great Western working to Portsmouth Harbour and an Arriva Trains Wales working to Manchester Piccadilly respectively. No. 158950 was formed by adding car No. 57751 to the complete No. 158761.

No. 158951, 1 October 2016
No. 158951 is seen on the buffer stops at Portsmouth Harbour, waiting to depart for Bristol Parkway. These services usually carry on to Cardiff Central, but due to engineering work inside the Severn Tunnel, they were terminated at Bristol Parkway. No. 158951 was formed of No. 158764 plus car No. 52751 from No. 158751.

No. 158952, 4 August 2018
No. 158952 stands at Cardiff Central waiting to depart for Portsmouth Harbour. There were twelve three-car Class 158/9 units in use with Great Western Railway, but half of them have now been reformed into their original sets, including No. 158952, which at the time was formed of No. 158762 with car No. 57745 added.

No. 158954, 3 July 2017

No. 158954 departs from Bristol Temple Meads with a Great Western Railway working through to Cardiff Central. This view shows off the former First Great Western local lines livery. Since this shot was taken, No. 158954 has reverted to its original formations, but at the time it was formed of No. 158760 with car No. 57747 added.

No. 158955, 3 July 2017

No. 158955 stands under the magnificent arch roof at Bristol Temple Meads carrying First Great Western local lines livery. This was formed of No. 158765 with car No. 52747 added, but this has since been disbanded, with car No. 52747 being removed and reformed with its sister car, and the remaining two cars becoming No. 158765 once more.

No. 158955, 18 July 2019
No. 158955 is seen arriving at Paignton station as it crosses over Torbay Road level crossing into the station. It would terminate at Paignton, then form the 12.53 Great Western Railway service to Cardiff Central.

No. 158956, 10 January 2001
No. 158956 is seen departing from Newport with a Central Trains working through to Cardiff Central. There were just eight three-car units in use with Central Trains at the time. They would be reformed back into two-car sets, then reformed again into three-car sets a few years later. No. 158956 was formed of No. 158793 with car No. 52784 added.

No. 158960, 3 July 2017
No. 158960 is seen at Bristol Temple Meads with a Great Western Railway working to Portsmouth Harbour. This unit was formed of No. 158769 with car No. 57749 added. It has since been disbanded, with car No. 57749 being reformed back into No. 158749.

No. 158961, 7 September 2019
No. 158961 departs from Bristol Temple Meads with a Great Western Railway working to Portsmouth Harbour. No. 158961 was formed of No. 158767 with car No. 52749 added. It has since reverted to its original formations.

No. 159002, 17 March 2017

No. 159002 is seen departing from the carriage sidings at Clapham Junction as part of an empty stock move to London Waterloo. The unit carries South West Trains livery, and at the time it also carried the name *City of Salisbury*.

No. 159003, 6 June 2019

No. 159003 stands in the summer sunshine at Exeter St Davids, waiting to depart for London Waterloo. This unit now carries the livery of South Western Railway, who had taken over the running of the South West franchise from South West Trains in 2017. No. 159003 has lost its *Templecombe* name since it has been repainted.

No. 159004, 9 July 1993
No. 159004 passes through Clapham Junction when less than six months old, heading towards London Waterloo. The unit is operated by Network South East in this view, and would later be named *Basingstoke and Deane*.

No. 159005, 7 September 2019
No. 159005 stands at Exeter St Davids, having arrived with a terminating service from London Waterloo. The unit is seen carrying South Western Railway livery. Note how the large first-class area has been retained on the class following refurbishment.

No. 159008, 12 March 2015
No. 159008 slows for the station call at Clapham Junction. This was originally given the number '158880' but, as with the rest of the fleet, it was never carried. It was internally modified at Rosyth Dockyard before entering service with Network South East.

No. 159010, 18 August 1997
No. 159010 is seen shunting around the sidings at Clapham Junction. Note that the gangway fly doors are open in readiness for coupling to another class member. No. 159010 carries Network South East livery but with the South West Trains logos halfway along the bodyside.

No. 159011, 2 May 1999
No. 159011 departs from Yeovil Junction with a South West Trains working to Exeter St Davids. This unit differs from the previous photograph of No. 159010, as it only carries South West Trains lettering underneath the second saloon window, compared to the logo carried by No. 159010.

No. 159012, 2 March 2019
No. 159012 arrives at Clapham Junction carrying South Western Railway livery. The later repaints into this livery carried a more basic version when compared to that carried by No. 159003 – there are no bars on the livery; it is just plain.

No. 159013, 25 February 1993
No. 159013 is seen passing through Crewe on its way to Rosyth for internal modifications prior to entering service. It is seen coupled with Nos 159016 and 159017, and the units would spend around three months having the work carried out before heading back to Salisbury and an entry into revenue-earning service.

No. 159013, 1 August 2001
No. 159013 is seen at Plymouth after arriving with a terminating service. The unit carries former Network South East livery but with South West Trains logos. Also, unusually, it has a black front skirt instead of the more usual white.

No. 159014, 6 July 1999
No. 159014 is seen arriving at Exeter St Davids from the Plymouth direction. The unit still carries former Network South East livery, despite South West Trains having run the franchise since 1997.

No. 159014, 9 July 2016
No. 159014 again arrives at Exeter St Davids, this time over Station Road at the other end of the station. It now carries South West Trains livery.

No. 159016, 12 August 2017
No. 159016 is seen departing from Basingstoke while heading for Exeter St Davids. It is seen working in multiple with classmate No. 159022, with both carrying South West Trains livery.

No. 159018, 1 April 1995
No. 159018 stands at Paignton when just a couple of years old. The unit is still in as-delivered condition, carrying Network South East livery. Note the Regional Railways branding on the station building.

No. 159018, 14 April 2016
No. 159018 departs from the very impressive Bristol Temple Meads station with a South West Trains working to London Waterloo. South West Trains had just over one more year of running these services before the franchise changed hands to South Western Railway.

No. 159021, 21 September 1999
No. 159021 is seen at Salisbury alongside Class 158 No. 158824. Salisbury has been the home depot for the Class 159 units since their introduction into service in 1993.

No. 159021, 5 April 2002
No. 159021 is seen at Paignton having just been repainted from Network South East livery into South West Trains livery. The unit carries a smaller than usual South West Trains logo above the first saloon window, which would soon be changed to the more usual style and size.

No. 159022, 17 May 1998
No. 159022 is seen on display at Bournemouth open day in 1998. The unit carries the running number '159023' instead of its correct number of '159022'. It also carries South West Trains logos halfway along the bodyside, and its carriage number and set number together under the first saloon window.

No. 159022, 12 August 2017
No. 159022 makes a station call at Basingstoke with a South West Trains working through to Exeter St Davids, working in multiple with classmate No. 159016.

No. 159102, 2 March 2019
No. 159102 is seen slowing for the station call at Clapham Junction. This unit carries former South West Trains livery but with the addition of orange flashes; however, it also carries South Western Railway logos on the bodyside. No. 159102 was rebuilt from Class 158 No. 158803 in 2006.

No. 159104, 3 July 2017
No. 159104 departs from Bristol Temple Meads carrying another livery variation, this time a different version of the coloured stripes, and also the addition of a different style of South West Trains logo. This unit was rebuilt from Class 158 No. 158805.

No. 159105, 2 April 2009
No. 159105 is seen at Salisbury carrying South West Trains livery. This was rebuilt from Class 158 No. 158807. There were eight Class 159/1 rebuilds in 2006 to compliment the twenty-two Class 159/0 units.